JUN 21

FARMER LLAMA'S FARM MACHINES

SPREADERS

BY KIRSTY HOLMES

Minneapolis, Minnesota

Library of Congress Cataloging-in-Publication Data is available at www.loc.gov or upon request from the publisher.

ISBN: 978-1-64747-546-8 (hardcover)
ISBN: 978-1-64747-553-6 (paperback)
ISBN: 978-1-64747-560-4 (ebook)

© 2021 Booklife Publishing
This edition is published by arrangement with Booklife Publishing.

North American adaptations © 2021 Bearport Publishing Company. All rights reserved. No part of this publication may be reproduced in whole or in part, stored in any retrieval system, or transmitted in any form or by any means, electronic, mechanical, photocopying, recording, or otherwise, without written permission from the publisher.

For more information, write to Bearport Publishing, 5357 Penn Avenue South, Minneapolis, MN 55419. Printed in the United States of America.

IMAGE CREDITS

All images are courtesy of Shutterstock.com, unless otherwise specified. With thanks to Getty Images, Thinkstock Photo, and iStockphoto. Cover – NotionPic, Tartila, A-R-T, logika600, BiterBig, Hennadii H. Aggie – NotionPic, Tartila. Grid – BiterBig. Farm – Faber14. Spreaders – Hennadii H. 2 – Fancy Tapis. 4 – Fancy Tapis. 5 – Mascha Tace. 6 – Colorcocktail. 7 – Dendi Ismi Sofyan, Fancy Tapis. 8 – Janis Abolins, Lunarus, Marina Akinina. 9 – Fancy Tapis, azar nasib, Tarikdiz, venimo, Tarikdiz, Lukas Beno. 11 – GoodSeller, 12&13 – BlueRingMedia. 13 – Ira Yapanda, Marina Akinina, K-Nick, yafi4. 14&15 – Fancy Tapis. 16 – Flat vectors. 17 – Scharfsinn, Suwin. 20 – Mascha Tace. 21 – DRogatnev. 22 – Farah Sadikhova, Visual Gen, Piyawat Nandeenopparit eration, Alexandr III. 23 – Fancy Tapis, Miloje, Viktor96.

CONTENTS

Down on the Farm! 4
What Is a Spreader? 6
Marvelous Manure 8
Types of Spreaders 10
Feed the Earth 12
And Feed the World! 14
A Job for a Spreader 16
Record Breakers 18
Get Your Llama-Diploma 20
The Need for Speed 22
Glossary 24
Index 24

DOWN ON THE FARM!

Welcome to Happy Valley Farm. I'm Aggie, and I'm a farmer llama.

Pee-yew! She got the spreader out again, didn't she?

Don't worry about that smell. Once you have been through your training, you will know what that is. Let's get started!

WHAT IS A SPREADER?

As **crops** grow in fields, they use up **nutrients** in the soil. A spreader is a machine that spreads, sprays, or scatters **fertilizer** onto a field to put nutrients back.

Liquid fertilizer can be sprayed onto the soil.

Many things can be used as fertilizer.

Compost is made of rotted foods and can be mixed with soil.

Fertilizer can be in small **pellets** that are scattered.

Compost is Aggie's favorite!

MARVELOUS MANURE

Chicken manure is great for vegetable gardens.

Cow manure is soft and easy to spread across fields.

Spreaders add a special type of fertilizer to the soil. It's called manure! Manure is made up of **natural material**, such as poop and pee. It comes from animals.

Poop from dogs and cats can't be used as fertilizer. This is because poop from these animals can have bad **bacteria** and worms in it. You don't want those near your veggies!

Pigs can produce up to 13 pounds (6 kg) of poop every day!

There's hay and straw in manure, too!

TYPES OF SPREADERS

Different types of fertilizer need different machines to spread them across fields.

MANURE SPREADER

These spreaders have big blades, which chop the **solid** manure into little bits before adding it to the field.

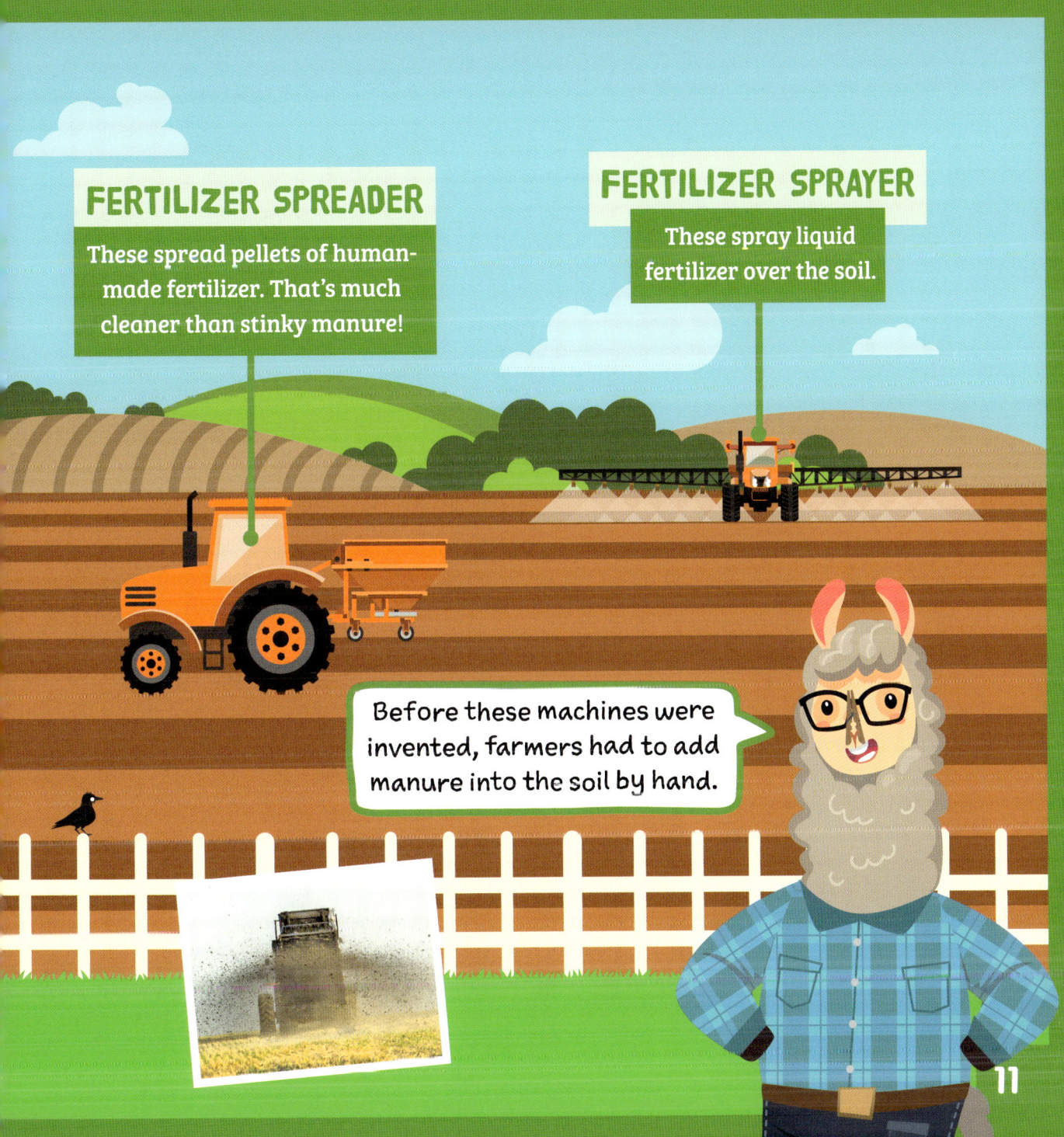

FEED THE EARTH

Farmers used to make sure soils kept their nutrients by letting fields rest. Each year, a different field would not have any crops planted in it at all. This gave time for the soil to recover.

FALLOW FIELD

The resting field is called the fallow field.

CROPS IN A FIELD

The other fields would have a different crop planted in them each year. Different crops take different amounts of nutrients from the soil.

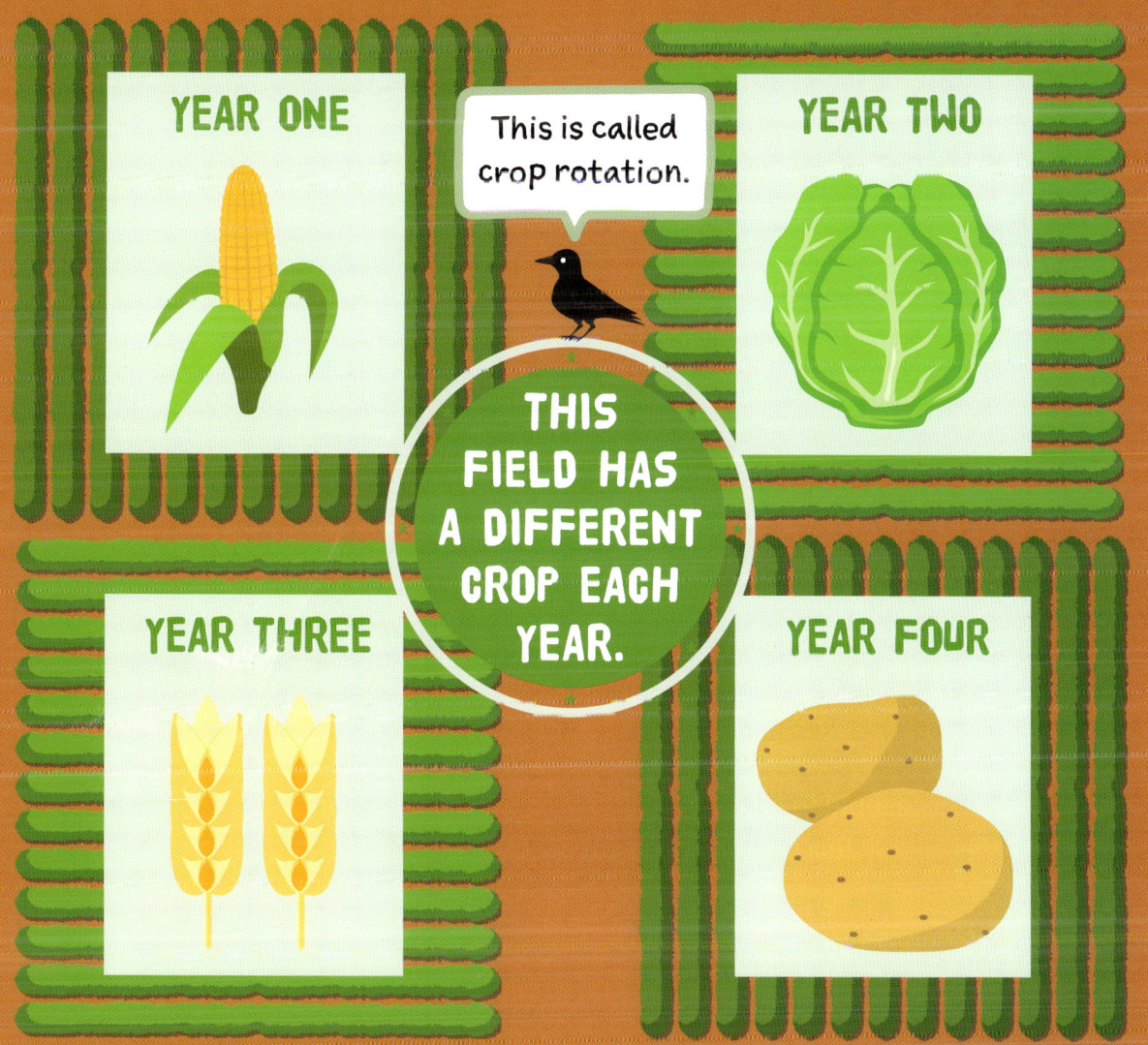

AND FEED THE WORLD!

With more than seven billion people on Earth, farmers have a lot of crops to grow. By fertilizing their soil, they might be able to use all their fields at once.

It is important that farmers make sure their soil has all the nutrients it needs. If they don't, the crops may not grow.

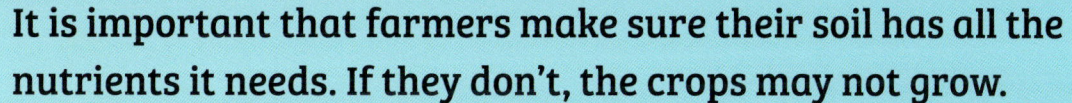

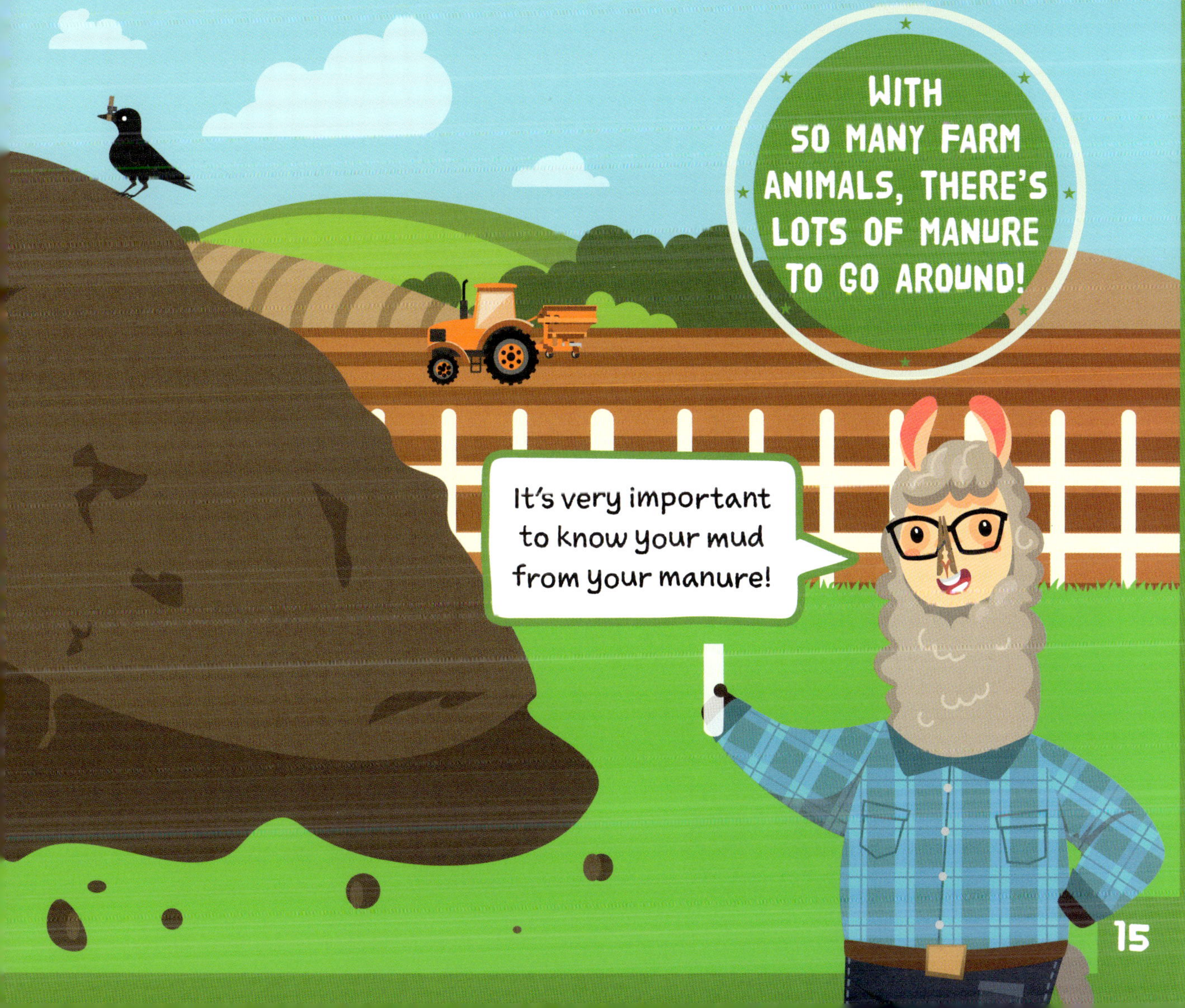

WITH SO MANY FARM ANIMALS, THERE'S LOTS OF MANURE TO GO AROUND!

It's very important to know your mud from your manure!

A JOB FOR A SPREADER

Spreaders can be used for other things, too. For example, when it is icy outside, a special spreader can be used to put salt on the roads. This helps the ice melt and makes the roads safe.

In the future, smart sprayers with computers in them might do this job for the farmer. Fertilizer could even be sprayed by **drones** flying over the field.

RECORD BREAKERS

MOST MANURE SPREAD

In 2014, a tractor and a manure spreader set the world record for the most manure spread in 24 hours. They spread over 9 million pounds (4 million kg) of manure.

I think this record stinks...

18

The record was set in Ukraine between July 31 and August 1, 2014. Even though they didn't all speak the same language, the farming team worked together to set this amazing record!

GET YOUR LLAMA-DIPLOMA

Well done! You made it through the training. If you've been paying attention, this test should be no prob-llama . . .

Questions

1. What is chicken manure good for?

2. What things from animals can be used as manure?

3. How is liquid fertilizer spread on a field?

4. What is a resting field with no crops called?

5. What can spreaders put on icy roads to make them safe?

You made that look easy! Welcome to the Happy Valley Farm family!

Answers: 1. Vegetable gardens 2. Poop and pee 3. It is sprayed on 4. A fallow field 5. Salt

Download your llama-diploma!

1. Go to **www.factsurfer.com**

2. Enter "**Spreaders**" into the search box.

3. Click on the cover of this book to see the available download.

21

THE NEED FOR SPEED

There must be a faster way to spread this manure. I feel the need . . . the need for speed!

STEP ONE — Manure

STEP TWO — Spreader

STEP THREE — Fast car